ADVENTURE
THE STORY OF ISRO

By

Govardhan Yepuri

Adventure The Story Of ISRO

by

Govardhan Yepuri

email: **govardhanyepuri7@gmail.com**

ISBN (Paperback): 978-81-987719-9-5

First Edition
March 2026

Print On Demand

Published By: Kasturi Vijayam
Ph:0091-9515054998
Email: Kasturivijayam@gmail.com

Book Available
@
Amazon (WorldWide), flipkart

Tommarow is my exam but I don't care because.... a single sheet of paper cannot decide my future….

A.P.J.Abdul Kalam

THE STORY OF ISRO

Introduction

The Indian Space Research Organisation (ISRO) is India's national space agency, founded in 1969 to promote the use of space technology for national development. Headquartered in Bengaluru, it works under the Department of Space. ISRO's main aim is to use space science for communication, weather forecasting, navigation, disaster management, and agriculture. It has developed many satellites and launch vehicles to support these activities. ISRO is known for its cost-effective and successful missions, such as Chandrayaan for lunar exploration and Mangalyaan for Mars. Through continuous research and innovation, ISRO plays a major role in advancing India's space capabilities and contributing to global space exploration.

The Adventures of ISRO show how dreams, dedication, and courage can turn a developing nation into a space power.

ISRO's journey is not just science — it's an inspiring story of India itself, proving that no dream is too big if we dare to chase it.

"Just as ISRO dared to dream of reaching the stars with limited resources, you too can achieve greatness by believing in your vision and persevering through challenges. Your journey may start small, but with passion and commitment, you have the power to transform dreams into realities that impact the world."

PART -1

The Beginning of ISRO

The Early Days of ISRO

DR. VIKRAM SARABHAI – (The Visionary Founder)

Dr. Vikram Sarabhai is known as the Father of the Indian Space Programme. He believed that space technology should help solve India's real problems. He founded ISRO in 1969, started the first rocket launching station at Thumba, and began using satellites for communication, education, and weather forecasting.

After India gained independence in 1947, a young scientist named Vikram Sarabhai (1919–1971) began researching cosmic rays. Influenced by C.V. Raman, a Nobel Prize-winning physicist, Sarabhai set up a laboratory to explore this area of science. His work soon expanded toward the use of space technology for national development.

Another eminent scientist, Homi J. Bhabha (1909–1966), had also studied cosmic rays during the 1940s. Sarabhai recognized that satellites could be useful tools for the progress of developing countries. He took the initiative to create institutions and frameworks that would make space research possible in India.

To organize and guide these efforts, the Indian National Committee for Space Research (INCOSPAR) was established in 1962. This committee was responsible for setting the goals and direction of India's early space program, laying the foundation for what would later become ISRO.

Interest in rocket technology began to grow when Thumba, near Thiruvananthapuram, was selected as a global site for launching sounding rockets because it lies close to the magnetic equator. The first foreign rocket was launched from Thumba in 1963, marking the start of India's space research activities. Since then, several countries have used the Thumba range for rocket launches under the United Nations program.

The Indian Space Research Organisation (ISRO) started a plan to make its own sounding rockets and the fuel needed for them. For this, ISRO built a rocket-making centre, a fuel plant, and a testing complex at Thumba in Kerala. A Space Science and Technology Centre was also set up there to do research and experiments. On the east coast, the island of Sriharikota was chosen as a new rocket launch centre.

In Ahmedabad, ISRO opened another centre to use space technology in different fields. After Dr. Vikram Sarabhai passed away in 1971, the Thumba centre was renamed the Vikram Sarabhai Space Centre (VSSC) to honour him. Later, in 1972, a satellite centre was started in Bangalore to design, build, and launch satellites.ISRO has many centres to help with launching and tracking rockets. The main ones are in Sriharikota, Thumba, and Balasore in Odisha. The control centre for all these is in Bangalore.

There are also tracking stations in Lucknow, Car Nicobar Islands, and Mauritius to follow the rockets and satellites. Another important optical tracking station is at Kavalur in Tamil Nadu.

Major contribution by Indian scientists

1.Dr. A.P.J ABDUL KALAM —(The Missile Man Of India)

("To succeed in your mission, you must have single-minded devotion to your goal".)

Dr. A.P.J. Abdul Kalam made significant contributions to the Indian Space Research Organisation (ISRO) during his early career, playing a key role in shaping India's space program. He joined ISRO in the 1960s and was a crucial part of India's first satellite launch vehicle project, the SLV-3 (Satellite Launch Vehicle-3), which successfully placed the Rohini Satellite into orbit in 1980. This achievement made India a member of the exclusive group of nations capable of launching satellites indigenously. As the Project Director of SLV-3, Dr. Kalam led a team of scientists and engineers, overcoming numerous challenges to make the mission a success. His leadership, vision, and dedication laid the foundation for the development of more advanced launch vehicles like the ASLV and PSLV. Dr. Kalam's work at ISRO not only strengthened India's technological capabilities but also inspired a generation of scientists to pursue excellence in space research and development.

2.Dr. SATHISH DAWAN — (The Organiser and Leader)

("We failed! But I have very strong trust in my team and I believe that next time we will definitely succeed")

Dr. Satish Dhawan made outstanding contributions to the growth and success of the Indian Space Research Organisation (ISRO) and is often called the "Architect of the Indian Space Programme." He became the Chairman of ISRO in 1972, succeeding Dr. Vikram Sarabhai, and guided the organization during a crucial period of development. Under his leadership, ISRO achieved major milestones such as the successful launch of India's first satellite, Aryabhata (1975), and the development of the SLV (Satellite Launch Vehicle) and Bhaskara remote sensing satellites. Dr. Dhawan strongly emphasized the importance of indigenous technology, self-reliance, and teamwork. He also reorganized ISRO to focus on practical applications of space technology for communication, meteorology, and remote sensing. Known for his humility and leadership, Dhawan encouraged scientific freedom and innovation among his team members, including Dr. A.P.J. Abdul Kalam. His vision and guidance transformed ISRO into one of the world's leading space agencies, and in his honor, India's main space center at Sriharikota was named the Satish Dhawan Space Centre (SDSC).

3.Dr. Udupi Rama Chandra Rao – (The Satellite Pioneer)

He developed India's first satellite Aryabhata (1975) and led the creation of INSAT and IRS satellites. These helped India grow in communication, remote sensing, and television broadcasting.

4.Dr. K. Kasturirangan – (The Scientific Visionary)

He expanded ISRO's satellite missions and initiated the idea of Chandrayaan-1, India's first mission to the Moon. He also guided the early PSLV and GSLV programs.

5.Dr. G. Madhavan Nair – (The Technologist)

He led the Chandrayaan-1 mission, which discovered water on the Moon. He also developed GSLV rockets and improved India's cryogenic engine technology.

6.Dr. K. Radhakrishnan – (The Mars Mission Leader)

Under his leadership, ISRO achieved the historic Mars Orbiter Mission (Mangalyaan) in 2013. India became the first country to reach Mars on its first attempt and at a very low cost.

7. Dr. K. Sivan – (The Modern Rocket Man)

He developed the GSLV Mk-III, led Chandrayaan-2, and prepared India for its first human spaceflight Gaganyaan. He is known for his simplicity and dedication.

8. Dr. S. Somanath – (The New Era Leader)

Dr. S. Somanath guided Chandrayaan-3 to a successful landing near the Moon's south pole in 2023. He is leading new missions like Aditya-L1 (Sun mission) andGaganyaan.

Dr. Nambi Narayanan–(The Unsung Hero of India's Cryogenic Dream)

Dr. Nambi Narayanan is a former scientist and aerospace engineer at ISRO, known for his pioneering work in developing cryogenic rocket engine technology — a crucial system for launching heavy satellites into space. Born on December 12, 1941, in Nagercoil, Tamil Nadu, he joined ISRO in the 1960s and played a key role in advancing India's space propulsion systems.

He led the team that worked on liquid propulsion when most of the world was still using solid fuel rockets. Under his guidance, ISRO successfully developed engines like Vikas, which became the backbone of India's PSLV and GSLV launch vehicles.

In 1994, Dr. Narayanan was falsely accused of espionage, which led to his arrest and a long legal battle. However, after years of struggle, he was completely exonerated by the Supreme Court of India in 1998. Despite this hardship, he continued to inspire scientists and youth with his dedication and patriotism.

In 2019, the Government of India honored him with the Padma Bhushan, recognizing his immense contributions to India's space research and his unbroken spirit. His story is a symbol of courage, perseverance, and truth triumphing over injustice.

Types of Launch Vehicle

The launch vehicles serve multiple purposes, such as sending communication, weather, navigation, and scientific research satellites into space. Examples include India's PSLV (Polar Satellite Launch Vehicle), GSLV (Geosynchronous Satellite Launch Vehicle), and SSLV (Small Satellite Launch Vehicle) each designed with unique capacities and technologies to handle different types of missions.

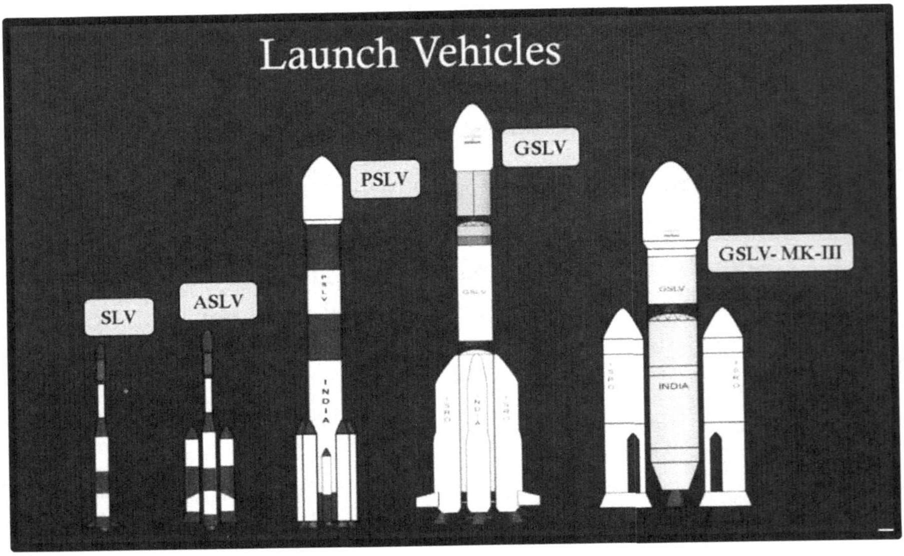

1.Satellite Launch Vehicle

The SLV program by ISRO was India's first experimental satellite launch vehicle, developed to place small satellites into Low Earth Orbit (LEO). It was introduced by A.P.J Abdul Kalam. RH-100, RH-125, RH-300, and RH-560 are launched using Satellite Launch Vehicle (SLV).

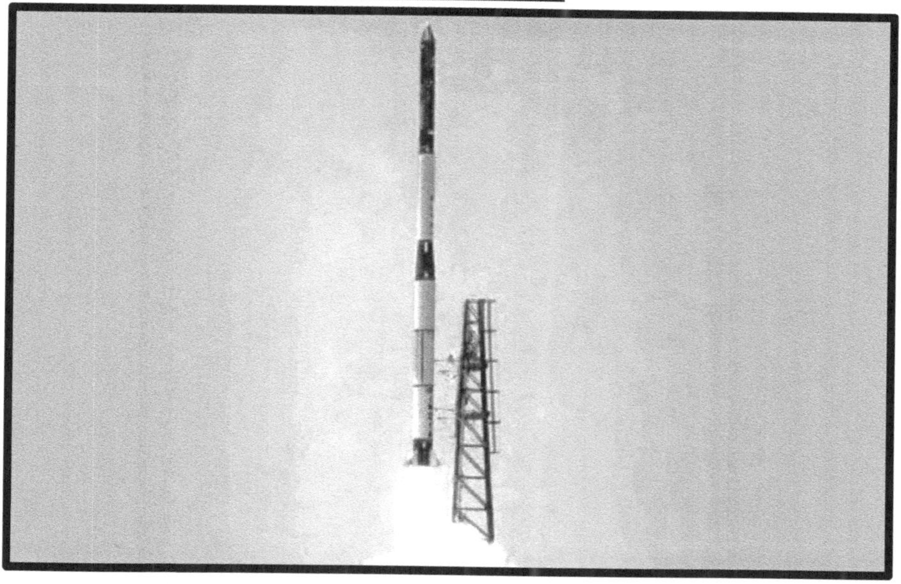

2. Augmented Satellite Launch Vehicle

The Augmented Satellite Launch Vehicle (ASLV) was an Indian launch vehicle developed by the Indian Space Research Organisation (ISRO) to serve as an intermediate step between the SLV and the more advanced launch systems. Designed to carry heavier payloads of up to 150 kg into Low Earth Orbit (LEO), the ASLV marked a significant technological advancement for India's space program. It was a five-stage, all-solid propellant rocket equipped with two strap-on boosters on its first stage. Operational between 1987 and 1994, the ASLV played a crucial role in developing key technologies such as closed-loop guidance, strap-on booster configuration, and inertial navigation systems, which later contributed to the success of ISRO's PSLV and GSLV programs. Its purpose was to provide a low-cost and efficient launch option with greater payload capacity than the earlier SLV.

3.Polar Satellite Launch Vehicle

The Polar Satellite Launch Vehicle (PSLV) is a launch vehicle developed by the Indian Space Research Organisation (ISRO). It was first launched on September 20, 1993, from the Satish Dhawan Space Centre, Sriharikota. PSLV is known for its ability to place satellites into different orbits such as polar, sun-synchronous, and geostationary transfer orbits. The rocket operates in four stages, using a combination of solid and liquid fuels, which makes it both powerful and reliable. Often called ISRO's "Workhorse Launch Vehicle," PSLV has been used to carry multiple satellites in a single mission and offers a cost-effective way to access space.

Some of its major achievements include launching Chandrayaan-1 (India's first lunar mission), Mangalyaan (India's first Mars mission), and setting a world record in 2017 by launching 104 satellites at once. PSLV has several versions such as PSLV-G, PSLV-CA, and PSLV-XL, designed for different mission needs. This rocket has played a vital role in making India self-reliant in space technology and has strengthened the country's position in the global space community.

4. Geosynchronous Satellite Launch Vehicle

The Geosynchronous Satellite Launch Vehicle (GSLV) is an Indian rocket developed by ISRO to launch heavy satellites into high orbits such as the Geosynchronous Transfer Orbit. It was first launched on April 18, 2001, from the Satish Dhawan Space Centre in Sriharikota. The GSLV has three stages — a solid-fuel first stage, a liquid-fuel second stage, and a cryogenic third stage that uses liquid hydrogen and liquid oxygen. It can carry communication, weather, and navigation satellites. The improved version, GSLV Mk III (also called LVM3), is India's most powerful rocket and was used for missions like Chandrayaan-2 and Chandrayaan-3. This launch vehicle shows India's progress in space technology and helps the country launch heavy satellites without depending on foreign rockets.

Early Satellite / Launch Capability

- 1975 – Aryabhata India's first satellite. Marked India's entry into space.
- 1980 – SLV-3 / Rohini First indigenous launch vehicle that put a satellite (Rohini) into low Earth orbit. Satellite Systems for Communication / Earth Observation.
- INSAT series A large communication and meteorological satellite system. Helped with broadcasting, weather forecasting, tele-education, telemedicine etc.
- IRS / Remote Sensing satellites.Enhanced imaging for agriculture, resource mapping, disaster management. E.g. IRS-1B, IRS-P2 etc.
- RISAT-1 (2012) First radar imaging satellite of India (SAR). Could image through clouds, useful in all-weather surveillance etc.

Navigation / Regional System.

- NavIC / IRNSS India's regional navigation satellite system. Provides positioning services over India and nearby regions. Reduced dependence on foreign systems.

Interplanetary / Lunar / Solar Missions

- 2008 – Chandrayaan-1 First mission to the Moon. Discovered water molecules on the lunar surface.
- 2013 – Mars Orbiter Mission / Mangalyaan. First Asian country to reach Mars orbit, and that too on first try. Low-cost mission, big prestige.
- 2019/2020 – Chandrayaan-2 Orbiter was successful; lander attempt failed but still many accomplishments in science and technology.
- 2023 – Chandrayaan-3 Soft-landing near Moon's south pole. India became first country to achieve a soft landing in that region.
- 2023 – Aditya-L1 First solar observatory mission; placed at L1, to continuously observe the Sun (photosphere, corona etc.).

Launch Vehicle / Propulsion / Technology Advances

- SLV family, Cryogenic Engines Development of vehicles capable of placing heavy satellites in geostationary orbit; indigenous cryogenic upper stage is a technical milestone.
- Reusable Launch Vehicle (RLV) tests / Pushpak etc. Demonstrating landing / recovery technologies, moving toward reusable technology.
- Record satellite launches / rideshare missions. E.g. PSLV-C37 mission that launched 104 satellites in one go. This showed ISRO's capability in small satellite launches & international rideshare.

Recent / Miscellaneous Milestones

- XPoSat (2024) India's X-ray Polarimeter Satellite; one of very few in world doing X-ray polarimetry.
- INSAT-3DS (2024) Weather, environmental monitoring improvements.
- SpaDeX mission – satellite docking etc. ISRO achieved satellite docking and undocking (space docking) which is crucial for future space.

Major satellites launched by ISRO

The Indian Space Research Organisation (ISRO) has launched many important satellites for communication, remote sensing, navigation, and space exploration. These satellites have helped India progress in science, technology, and national development.

1. Aryabhata (1975)

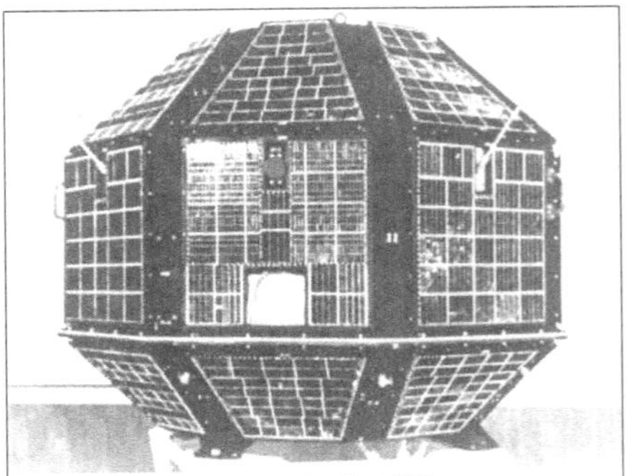

Aryabhata Satellite (1975)

Aryabhata was India's first satellite, launched on April 19, 1975, from Russia (then USSR). It was named after the great Indian astronomer Aryabhata. The satellite was designed to conduct experiments in X-ray astronomy and solar physics. Even though it remained operational for only a short time, it marked the beginning of India's journey into space research, laying the foundation for future ISRO missions.

2. Rohini Series (1979–1994)

The Rohini series of satellites were India's first satellites launched using Indian-made rockets. The first satellite, Rohini-1, was launched in 1979 but was unsuccessful. However, Rohini Satellite RS-1, launched in 1980 by the SLV-3 rocket, became the first satellite launched from Indian soil. The Rohini satellites were mainly used for scientific experiments and Earth observation, and they demonstrated India's growing capability to design, build, and launch satellites independently.

3. Kalpana-1 (2002)

Kalpana-1, launched on September 12, 2002, was a dedicated meteorological satellite named in honor of Kalpana Chawla, the first Indian-born woman astronaut. It was designed to monitor weather patterns, track cyclones, and collect meteorological data. The satellite played an important role in weather forecasting and disaster management, helping India predict and respond to natural calamities more effectively.

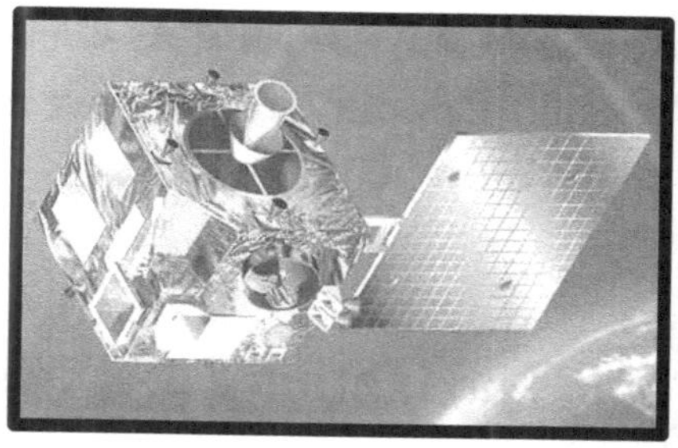

4. INSAT and GSAT Series

The INSAT (Indian National Satellite) and GSAT (Geostationary Satellite) series are among India's most important communication satellites. These satellites provide services for television broadcasting, telecommunication, meteorology, and telemedicine. For example, INSAT-1B (1983) improved India's TV and telephone networks, GSAT-6A (2018) offered secure communication for defense purposes, and GSAT-30 (2020) replaced INSAT-4A to continue broadcasting and communication services. Together, these satellites have made India largely self-reliant in communication technology.

5. Earth Observation Satellites

ISRO's Earth Observation Satellites are used to study the Earth's surface and environment. The IRS (Indian Remote Sensing) satellites, launched since 1988, help in agriculture, forestry, and water resource management. The Cartosat series provides high-resolution images useful for urban planning and mapping, while the RISAT (Radar Imaging Satellite) series offers day-and-night, all-weather imaging capabilities. These satellites play a vital role in disaster management and national development.

6. Navigation Satellites (IRNSS / NavIC)

The Indian Regional Navigation Satellite System (IRNSS), also known as NavIC (Navigation with Indian Constellation), was launched between 2013 and 2018. This system provides accurate positioning and timing services across India and nearby regions, similar to the American GPS. NavIC is used for navigation in land, air, and sea transport, as well as for mapping and personal location services.

7. Space Exploration Missions

ISRO has also achieved great success in space exploration missions. Chandrayaan-1 (2008), India's first lunar mission, discovered water molecules on the Moon's surface. Chandrayaan-2 (2019) aimed to explore the Moon further, with its orbiter still operational today. Chandrayaan-3 (2023) made history by achieving a successful soft landing on the Moon's south pole. The Mars Orbiter Mission (Mangalyaan, 2013) made India the first Asian nation to reach Mars orbit, and

Aditya-L1 (2023) became India's first solar mission to study the Sun from space. These missions have placed India among the top spacefaring nations of the world.

8. Student and Small Satellites

ISRO also encourages innovation through student and small satellite projects. Anusat (2009) was the first university-built satellite developed by Anna University, Chennai. YouthSat (2011) was a joint Indo-Russian mission that studied the effects of solar radiation on the Earth's atmosphere. Microsat-R (2019) was a small satellite used for imaging and research purposes. These missions help young scientists gain practical experience in satellite technology and contribute to India's growing space capabilities.

"From Aryabhata to Aditya-L1, ISRO's satellites have made India a global space leader. These missions have improved communication, weather forecasting, navigation, and space exploration, showing India's technological excellence and vision for the future."

****

PART-2

Major Missions

ROHINI –75

The Rohini 75 (RH-75) was India's first indigenously developed sounding rocket, marking a major milestone in the nation's space journey. It was successfully launched on November 20, 1967, from the Thumba Equatorial Rocket Launching Station (TERLS) in Kerala, located near the Earth's magnetic equator. With a diameter of 75 mm and a total weight of about 32 kg, the RH-75 was designed to carry scientific instruments for atmospheric and space research during its sub-orbital flight. Its development represented the foundation of India's capability to design and build rockets independently. The success of the Rohini 75 paved the way for more advanced rockets in the Rohini series and contributed significantly to the evolution of ISRO's larger launch vehicles, establishing the groundwork for India's future space exploration achievements.

After the success of Rohini-75 then the series expanded. The diameter of the rockets steadily increased along with their sophistication. RH-100, RH-125, RH-300, and RH-560 are launched using Satellite Launch Vehicle (SLV).

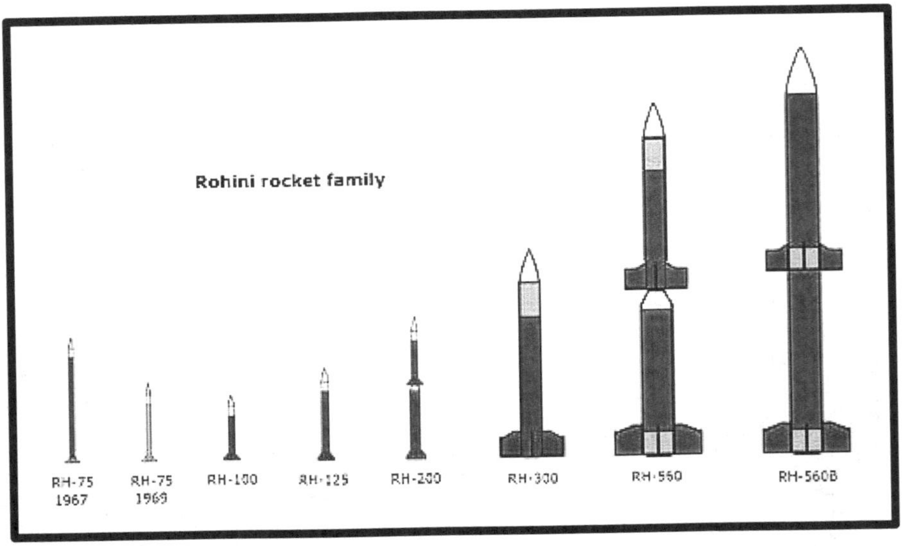

MANGALYAAN

The Mars Orbiter Mission (MOM), popularly called Mangalyaan, marks a historic milestone in India's space exploration journey. Launched on November 5, 2013, by the Polar Satellite Launch Vehicle (PSLV-C25) from Sriharikota, it represented India's first attempt to reach another planet. After traveling through space for nearly 300 days, the spacecraft successfully entered Mars' orbit on September 24, 2014, a remarkable feat that placed India among the world's leading spacefaring nations.

This mission had two main objectives technological and scientific. On the technological side, it tested and validated complex systems required for deep space communication, navigation, and orbit control during the long interplanetary journey. Scientifically, Mangalyaan carried instruments designed to study Mars' surface, atmosphere, and mineral composition, providing valuable data on the planet's climate and terrain. The mission also helped Indian scientists gain experience in interplanetary spacecraft design and operations.

Mangalyaan's success brought global recognition to the Indian Space Research Organisation (ISRO). India became the first Asian nation to reach Mars and the first country in the world to do so on its maiden attempt. Though it was originally planned to last six months, the mission continued sending data for more than seven years, far exceeding its expected lifespan. Beyond its scientific and technological achievements, Mangalyaan symbolized India's growing capability in space research and inspired millions across the nation.

CHANDRAYAAN-1

Chandrayaan-1 was India's first mission to explore the Moon, launched by ISRO on October 22, 2008, from the Satish Dhawan Space Centre. The spacecraft orbited the Moon and carried several scientific instruments to study its surface and minerals. One of its most important discoveries was the presence of water molecules on the lunar surface. It also carried the Moon Impact Probe (MIP), which made a controlled impact on the Moon on November 14, 2008, marking India's first touch on the lunar surface. Even though contact with the spacecraft ended in August 2009, the mission was considered a major success and helped strengthen India's space exploration capabilities.

Chandrayaan-1 was launched by ISRO using the PSLV-C11 rocket, a powerful version of the Polar Satellite Launch Vehicle designed to send spacecraft beyond Earth's orbit. The rocket successfully placed the Chandrayaan-1 satellite into lunar orbit. The satellite weighed around 1,380 kilograms and carried 11 scientific instruments from India and several other countries. These instruments were used to study the Moon's surface, minerals, and atmosphere, leading to the major discovery of water molecules on the Moon.

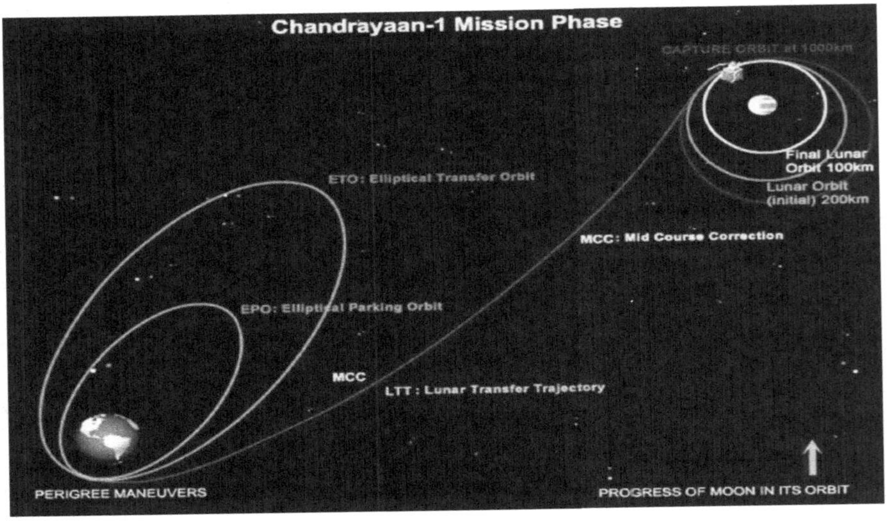

CHANDRAYAAN-2

Chandrayaan-2 was a lunar mission launched by the Indian Space Research Organisation (ISRO) on July 22, 2019, with the goal of exploring the Moon's south polar region. The mission consisted of three components an orbiter, lander, and rover. While the orbiter continues to function successfully and collect valuable scientific data, the lander Vikram lost communication during its descent in September 2019. The mission aimed to study the Moon's topography, seismography, mineral distribution, and its thin atmosphere.

Orbiter:

The orbiter successfully entered the Moon's orbit and carries eight scientific payloads. It continues to perform long-term observations and has exceeded its original one-year mission life, providing crucial data about the lunar surface and environment.

Vikram Lander:

Named after Dr. Vikram Sarabhai, the father of India's space program, the Vikram lander was designed to make a soft landing and operate for one lunar day (about 14 Earth days). Unfortunately, ISRO lost communication with the lander during its final descent on September 7, 2019, when it was just 2.1 km above the lunar surface.

Pragyan Rover:

The Pragyan rover, a six-wheeled robotic vehicle, was housed inside the Vikram lander. It was intended to perform experiments on the lunar surface, analyzing soil composition and studying the Moon's terrain.

The primary goals of Chandrayaan-2 were to advance India's capabilities in space exploration and deepen the understanding of the Moon. The mission aimed to develop and demonstrate new technologies required for future lunar missions. Another major objective was to perform a soft landing on the lunar surface and operate a rover for surface exploration. It also sought to study the Moon's topography and mineral composition, helping scientists learn more about its formation and evolution. Additionally, the mission aimed to map the distribution of water-ice in the polar regions and analyze the chemical composition and exosphere of the Moon to understand its atmosphere and surface interactions.

After a series of Earth-bound maneuvers, the spacecraft successfully performed Lunar Orbit Insertion on August 20, 2019. The Vikram lander separated from the orbiter on September 2, 2019, preparing for its descent. However, during the final phase of the landing on September 7, 2019, communication with the lander was lost at an altitude of about 2.1 km from the lunar surface.

CHANDRAYAAN-3

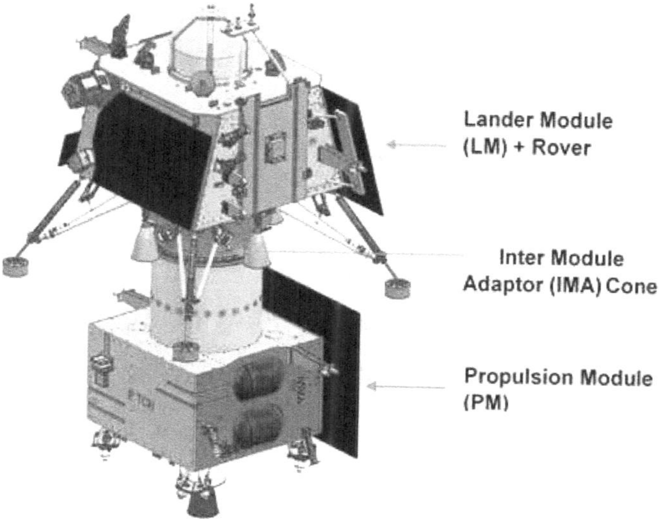

Chandrayaan-3 was India's third lunar exploration mission, launched by the Indian Space Research Organisation (ISRO) to demonstrate its ability to achieve a soft landing and operate a rover on the Moon's surface. The mission made history on August 23, 2023, when it successfully landed near the Moon's south pole an unexplored region known for its extreme conditions and potential presence of water ice. This achievement made India the fourth country to successfully perform a soft landing on the Moon and the first to reach the lunar south pole.

The mission consisted of three main modules: the Propulsion Module, the Lander (Vikram), and the Rover (Pragyan). The Propulsion Module was responsible for carrying the Lander and Rover to lunar orbit and also housed an experiment to study Earth's atmosphere. The Vikram Lander was designed to ensure a safe and precise soft landing, while the Pragyan Rover was deployed to conduct scientific investigations such as analyzing soil composition and detecting minerals on the lunar surface.

Chandrayaan-3 marked a major technological milestone for India's space program, showcasing advancements in landing systems, navigation, and rover operations. It not only strengthened India's position in global space exploration but also inspired the nation with its success after the Chandrayaan-2 setback in 2019. The mission's success proved ISRO's growing capabilities and set the foundation for future interplanetary missions, including possible lunar sample-return and Mars exploration projects.

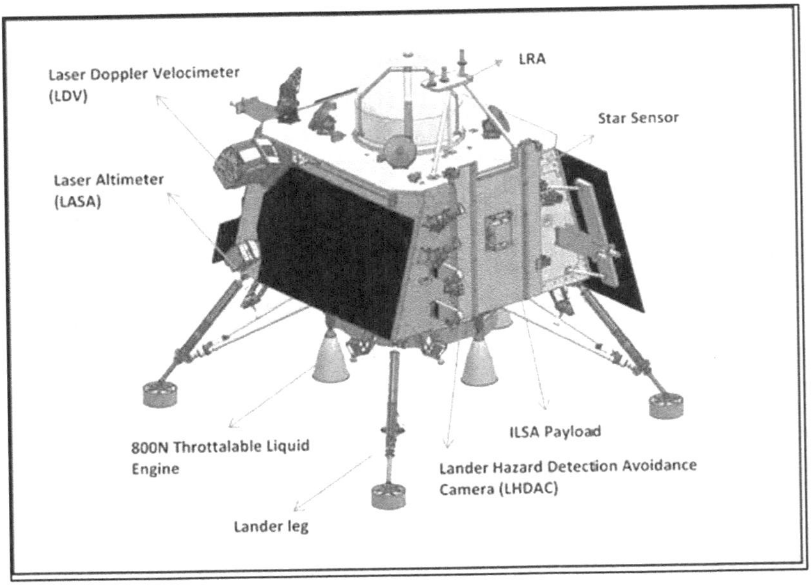

ADITYA- L1

Aditya-L1 is India's first dedicated solar mission, launched on September 2, 2023, by the PSLV-C57 rocket from the Satish Dhawan Space Centre. The mission aims to study the Sun from a unique vantage point, positioned in a halo orbit around the Sun-Earth Lagrange Point 1 (L1), located about 1.5 million kilometers from Earth. This position allows continuous, uninterrupted observation of the Sun without any eclipses or occultations. Aditya-L1 reached its designated orbit on January 6, 2024, after a several-month journey.

The mission carries seven scientific payloads—four for remote sensing of the Sun's photosphere, chromosphere, and corona, and three for in-situ studies of particles and fields around L1. These instruments help scientists investigate key solar phenomena such as the heating of the Sun's corona, the dynamics of solar flares, and Coronal Mass Ejections (CMEs) that influence space weather.

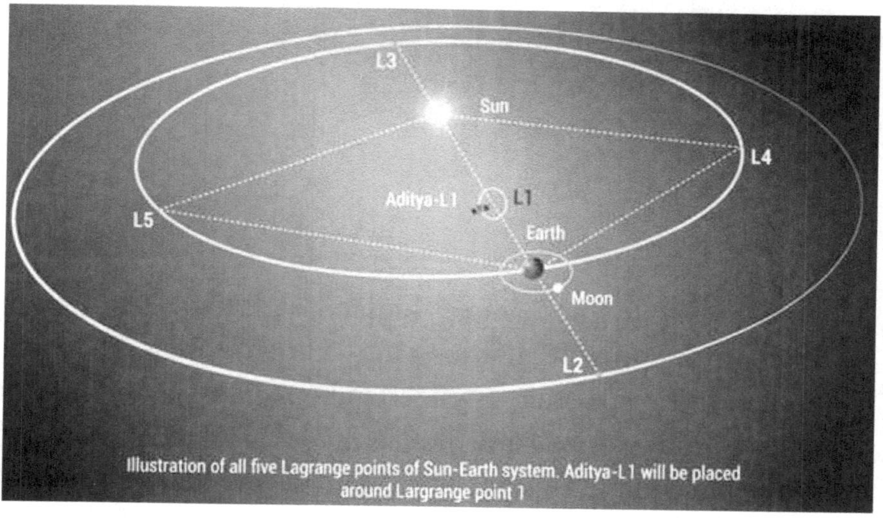

Illustration of all five Lagrange points of Sun-Earth system. Aditya-L1 will be placed around Largrange point 1

Among its notable instruments is the Solar Ultraviolet Imaging Telescope (SUIT), which has already captured the first full-disk ultraviolet images of the Sun, marking a major milestone in India's solar research journey.

104 SATELLITES ON SINGLE LAUNCH

On February 15, 2017, the Indian Space Research Organisation (ISRO) created a world record by successfully launching 104 satellites on a single rocket the Polar Satellite Launch Vehicle PSLV-C37 from the Satish Dhawan Space Centre, Sriharikota. The main payload of the mission was Cartosat-2D, a high-resolution Earth observation satellite developed in India for mapping, cartography, and urban planning purposes. Along with it, ISRO also launched two of its own nano-satellites, INS-1A and INS-1B, carrying scientific and experimental payloads from Indian academic institutions.

In addition to these three Indian satellites, 101 foreign satellites from six countries the United States, Israel, Netherlands, Kazakhstan, Switzerland, and the United Arab Emirates (UAE) were also deployed into orbit. The majority of these belonged to the United States, comprising 96 Dove Flock-3p nanosatellites developed by Planet Labs for global Earth imaging and environmental monitoring. The other international satellites included BGUSat from Israel for atmospheric studies, PEASSS from the Netherlands for technology testing, Al-Farabi-1 from Kazakhstan as an educational satellite,

SwissCube-2 from Switzerland for space research, and Nayif-1 from the UAE for amateur radio and educational purposes.

This incredible feat demonstrated ISRO's excellence in precision, cost-effectiveness, and reliability in the field of space technology. It marked a new milestone in the history of space exploration, showing India's ability to launch multiple satellites in a single mission. With the PSLV-C37 mission, ISRO not only strengthened India's global reputation in the space ommunity but also showcased its growing role in providing commercial satellite launch services for international customers.

Country	**No.of Satellite**	**Satellite Name**	**Purpose**
India	3	Cartosat-2D, INS-1A, INS-1B	Earth observation & experiments
USA	96	Dove (Flock-3p)	Global Earth imaging
Israel	1	BGUSat	Atmospheric research
Netherland s	1	PEASSS	Technology demonstration
Kazakhsta n	1	Al-Farabi-1	Educational
Switzerlan d	1	SwissCube-2	Research
UAE	1	Nayif-1	Amateurteur radio
Total	104		

GAGANYAAN

India stands on the threshold of a new dawn in its space journey a moment when dreams take flight beyond the skies and courage meets the cosmos. The Gaganyaan Mission is not just a technological milestone; it is a testament to the spirit, determination, and brilliance of countless Indian scientists who dare to push the boundaries of possibility. This historic mission aims to send three Indian astronauts into Low Earth Orbit (LEO), at an altitude of around 400 km, for a three-day mission, marking India's first step into human spaceflight. Using entirely indigenous technology, ISRO seeks to showcase the nation's ability to conduct independent human space missions, ensuring that the crew is launched, sustained in space, and brought safely back to Earth with a splashdown in the Indian Ocean.

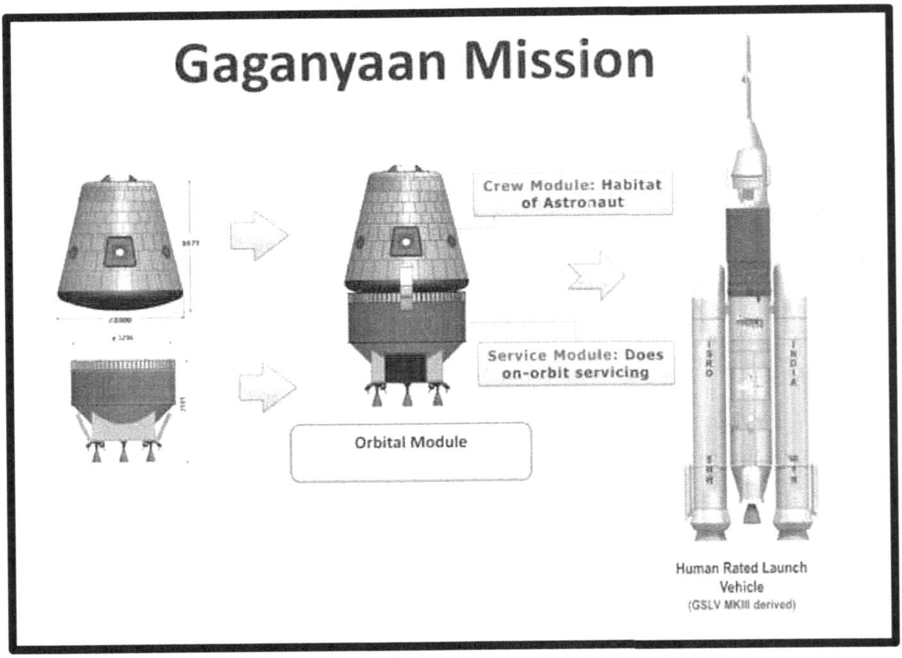

The mission is powered by the human-rated GSLV Mk III rocket, designed to carry the Crew Module (CM) — a secure, pressurized habitat for astronauts — and the Service Module (SM), which provides propulsion and power support. To ensure safety, a Crew Escape System (CES) has been developed to pull the crew away from danger in case of launch emergencies, while a Life Support System maintains a comfortable, Earth-like environment onboard. Before humans venture into space, Vyommitra, a humanoid robot, will undertake an unmanned mission to validate the spacecraft's systems and life-support capabilities.

The journey toward the stars is unfolding in carefully planned phases, including two unmanned test flights followed by the crewed mission, which is now targeted for 2027. With nearly 90% of the work already completed, ISRO's Gaganyaan represents not just a mission, but a dream — a promise that India's flag will soon shine among the stars, carried by the courage of its astronauts and the vision of a nation determined to reach beyond the horizon.

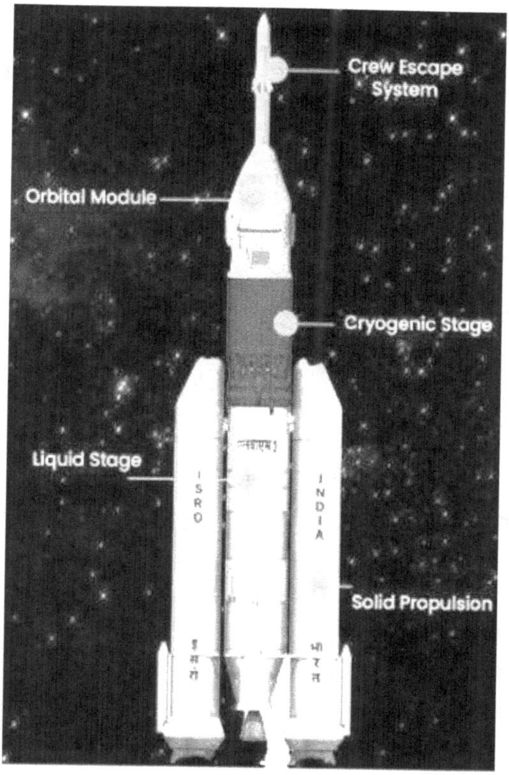

PART- 3
Global Collaborations
& Impact

International Collaborations and Impact

Bonding with NASA

The bonding between NASA (National Aeronautics and Space Administration, USA) and ISRO (Indian Space Research Organisation, India) is a remarkable example of international cooperation in space exploration. Their partnership reflects mutual respect, technological collaboration, and a shared vision of exploring space for the benefit of all humankind.

1. A Relationship Built on Trust and Mutual Growth

NASA and ISRO have been scientific partners for decades. Their collaboration began in the 1960s, when NASA helped India launch its first sounding rockets from the Thumba Equatorial Rocket Launching Station (TERLS). This early cooperation laid the foundation for India's future in space science and symbolized friendship between the two nations.

2. Major Collaborative Missions

a. Chandrayaan-1 (2008):

NASA contributed the Mini-SAR and M3 instruments, which helped confirm the presence of water molecules on the Moon a groundbreaking discovery.

This mission was a major success for both nations and deepened scientific ties.

b. NISAR (NASA–ISRO Synthetic Aperture RadarMission):

Set to launch soon, NISAR is a joint Earth-observing satellite that will study changes in the Earth's surface, including earthquakes, landslides, and climate change.

It is one of the most significant collaborations ever between the two agencies and uses both NASA's L-band radar and ISRO's S-band radar technology.

3. Human Spaceflight and Future Exploration

NASA is supporting ISRO's Gaganyaan Mission by sharing expertise in crew training, flight safety, and life-support systems.

Future cooperation may also extend to space station research, planetary exploration, and deep-space missions.

4. Data and Technology Exchange

Both agencies regularly share satellite data for monitoring natural disasters, environmental changes, and space weather.

Through programs like ISRO-NASA scientific workshops and training exchanges, scientists from both sides learn from each other's strengths ISRO's cost-effective engineering and NASA's advanced technology.

5. Symbol of Global Unity in Space

The NASA–ISRO bond represents more than scientific cooperation — it shows how nations can rise together beyond borders. Their teamwork highlights a belief that space belongs to humanity, not to one country alone.

From the shores of Thumba to the orbits of the Moon, NASA and ISRO's friendship has grown stronger with every mission. It's a partnership built on trust, curiosity, and the dream to explore the unknown together.

Their united vision reminds us that when minds across continents work as one, the sky is not the limit — it's just the beginning.

Bonding with ESA (European Space Agency)

The collaboration between ISRO (Indian Space Research Organisation) and ESA (European Space Agency) represents a strong partnership in space research, technology, and exploration. Their bond is built on mutual trust, scientific exchange, and the shared goal of advancing peaceful space exploration.

1. Early Collaborations

- The relationship began in the 1970s, when ESA helped India in launching early satellites and providing technical support.
- ESA launched India's Aryabhata satellite in 1975 and Bhaskara satellites in the late 1970s using European rockets.
- This cooperation laid the foundation for India's independent space launch capability later.

2. Joint Missions and Technology Sharing

- Chandrayaan-1 (2008) – ESA provided ground station support for tracking and data transmission through its facilities in Spain and Australia.
- Mars Orbiter Mission (2013) – ESA again assisted ISRO by providing deep space tracking via its ESTRACK network.
- This partnership ensured reliable communication with Indian spacecraft traveling far beyond Earth.

3. Current Cooperation

- ISRO and ESA continue to collaborate in Earth observation, space science, and deep space missions.
- Both agencies share satellite data, particularly for monitoring climate change, oceans, and natural disasters.
- Joint research on navigation systems (like Galileo and NavIC) also enhances global connectivity.

4. Future Prospects

The bond between ISRO and ESA is expected to strengthen with:

- Joint planetary missions (such as Venus and Mars).
- Space situational awareness cooperation to track space debris.
- Exchange programs for scientists and engineers to foster innovation.

5. Symbol of Global Unity

The ISRO–ESA relationship reflects how space agencies from different continents can unite for the betterment of humankind. Their teamwork proves that science knows no borders, and collaboration leads to discovery and progress for all.

Bonding with Russia (Former USSR)

The partnership between ISRO and Russia (formerly the Soviet Union) is one of the oldest and strongest in India's space history. This bond has shaped India's journey from its very first satellite to its future human spaceflight dreams.

1. The Beginning of the Bond

- The friendship began in the early 1970s, when India was still taking its first steps in space technology.
- The USSR helped India launch its first satellite, Aryabhata, on April 19, 1975, using the Kosmos-3M rocket from Kapustin Yar, Russia.
- This historic launch marked India's entry into space, creating a deep and lasting bond between the two nations.

2. Support in Satellite Launches

- Following Aryabhata, Bhaskara-I (1979) and Bhaskara-II (1981) were also launched by Soviet rockets.
- Several Indian Remote Sensing (IRS) and communication satellites were later launched with Russian assistance
- This collaboration allowed India to gain experience in

satellite design and operations long before it developed its own launch vehicles.

3.Human Spaceflight Connection

- In 1984, Rakesh Sharma became the first Indian in space, flying aboard the Soviet Soyuz T-11 spacecraft.
- His famous words — "Saare Jahan Se Achha" — from space became a proud symbol of India's unity and scientific spirit.
- This mission inspired generations of Indian scientists and laid the groundwork for India's future Gaganyaan program.

4. Technical and Scientific Cooperation

- Russia has helped ISRO with cryogenic engine technology, which is crucial for heavy-lift rockets like GSLV.
- Russian experts have shared knowledge on navigation systems, space medicine, and crew safety, all vital for upcoming human space missions.

5. Present and Future Collaboration

Today, India and Russia continue to work together on:

- Gaganyaan Mission: Russia provides training for Indian astronauts and space suits for the mission.
- Joint lunar and planetary research, including discussions for Luna and Venus missions.
- Space station cooperation and satellite navigation systems (GLONASS & NavIC) integration.

6. A Bond Beyond Technology

- The ISRO–Russia bond is not just about science — it is about trust, friendship, and shared dreams.
- From helping India take its first step into space to preparing Indian astronauts for space travel, Russia has stood as a true space partner.
- Their collaboration continues to symbolize the spirit of global cooperation for peace and exploration.

ISRO–France Collaboration

India and France share one of the oldest and strongest space partnerships in the world. Their cooperation began in the 1960s, soon after ISRO was founded, and continues successfully today through the Indian Space Research Organisation (ISRO) and the French Space Agency (CNES – Centre National d'Études Spatiales).

1.Satellite Development

- Megha-Tropiques (2011): A joint satellite mission to study tropical weather and climate.
- SARAL (2013): A mission for oceanographic studies, combining India's satellite expertise with France's ocean data technology.

2. Earth Observation and Climate Studies

- Both countries share satellite data to monitor climate change, ocean currents, and atmospheric conditions.
- They also work together in space-based applications for sustainable development.

3. Space Exploration and Future Missions

- India and France plan to cooperate on human spaceflight, especially for ISRO's Gaganyaan mission, where CNES will help train Indian astronauts and provide medical and life-support expertise.

4. Launch Cooperation

- India's PSLV rockets have launched several French satellites, showing strong mutual trust and technical coordination.

5. Maritime Surveillance

- ISRO and CNES have built joint satellites for tracking ships and maritime activities to enhance security and ocean management.

Technological Cooperation Between ISRO and DRDO

ISRO and DRDO work together to make India strong in space and defence. Their partnership is based on collaboration, not competition. ISRO mainly handles space technology like satellites and rockets, while DRDO focuses on defence systems such as missiles and military technology. Together, they help India become self-reliant and reduce dependence on foreign countries.

ISRO supports national security by providing satellites for defence purposes. These satellites help in border monitoring, surveillance, secure communication, and navigation through NavIC. They act like "eyes in space" for the Indian Armed Forces, giving important information for safety and strategic planning.

DRDO's missile programs benefit from ISRO's space expertise. Technologies related to rocket propulsion, guidance systems, materials, and avionics are shared between the two organizations. Because of this, missile systems like Agni and Prithvi have similarities with ISRO's launch vehicle technology.

A major example of their bonding is the Anti-Satellite (ASAT) test, Mission Shakti. In this mission, DRDO developed the interceptor missile, while ISRO provided satellite tracking and space data. This successful test proved India's capability to protect its space assets.

ISRO and DRDO also share knowledge, talent, and future goals. Scientists work together in research labs and exchange ideas in areas like AI, robotics, hypersonic technology, and space security. This strong cooperation creates a single scientific ecosystem that supports India's future defence and space missions.

"The bonding of ISRO and DRDO represents the fusion of space science and defence technology, strengthening India's strategic autonomy and national security."

PART - 4

Internships

&

Job Opportunities

ISRO Internship Programs

The Indian Space Research Organisation (ISRO) offers internship opportunities at its various centres across India for students pursuing engineering, science, and management disciplines. Centres such as the Vikram Sarabhai Space Centre (VSSC) in Thiruvananthapuram, the U R Rao Satellite Centre (URSC) in Bengaluru, the Liquid Propulsion Systems Centre (LPSC), and the Space Applications Centre (SAC) in Ahmedabad host interns regularly. Students pursuing B.E/B.Tech, M.E/M.Tech, or M.Sc in fields like Electronics, Computer Science, Mechanical, Aerospace, or Physics are eligible. These internships generally last for 4 to 8 weeks and allow students to work on real-time space technology projects. Applications are usually made through the official ISRO websites or via the student's Head of Department with a formal request letter from their institution.

IIRS (Indian Institute of Remote Sensing) Internship

The Indian Institute of Remote Sensing (IIRS) in Dehradun, a part of ISRO, offers internships and training programs in the fields of remote sensing, GIS, satellite image analysis, and space applications. The institute provides students with hands-on experience in using satellite data for environmental and developmental studies. It also offers e-learning opportunities through its EDUSAT-based online platform, allowing students from different parts of India to access high-quality space education remotely. This program is ideal for students interested in Earth observation and data interpretation technologies.

NRSC (National Remote Sensing Centre) Internship

The National Remote Sensing Centre (NRSC) in Hyderabad provides internships focusing on geospatial technologies, remote sensing, and satellite data processing. Students from Geoinformatics, Civil Engineering, Computer Science, and Earth Sciences backgrounds can participate in research and development projects related to mapping, agriculture, disaster management, and environmental monitoring. Through this internship, students gain valuable practical exposure to how satellite images are processed and used for solving real-world problems. The NRSC internship enhances both technical skills and understanding of India's space-based applications.

Young Scientist Programme (YUVIKA)

YUVIKA, also known as the "Young Scientist Programme," is an inspiring summer camp organized by ISRO for school students studying in Classes 9 and 10. The main objective of YUVIKA is to promote interest in space science and technology among young minds. Selected students visit ISRO centres, interact with scientists, and participate in hands-on learning sessions and experiments. The program also includes space awareness activities and model-making workshops, encouraging creativity and curiosity. It is one of ISRO's key initiatives to identify and nurture the next generation of space scientists and engineers.

RESPOND and Research Fellowship Programs

The RESPOND (Research Sponsored) and Fellowship Programs are designed for postgraduate and research-level students such as M.Tech, M.Sc, or Ph.D. candidates. These programs allow students to work on ongoing ISRO research projects in collaboration with their universities. ISRO also offers Junior Research Fellowships (JRF) and Research Associateships (RA) to support advanced scientific research in various space-related fields. Through these programs, students can contribute to innovations in satellite systems, propulsion, communication, and remote sensing technologies while being guided by experienced ISRO scientists.

Job Opportunities in ISRO

Scientific and Technical Positions

The scientific and technical roles form the backbone of ISRO. These professionals design, develop, and test rockets, satellites, and space systems that drive India's space missions. The most common entry-level position is Scientist/Engineer – SC, along with other posts like Technical Assistant, Technician, Research Scientist, and Junior Research Fellow (JRF). Candidates are recruited mainly through the ISRO Centralised Recruitment Board (ICRB), which conducts a written examination followed by an interview.

Applicants must hold a B.E./B.Tech degree in branches such as Electronics and Communication, Mechanical, Computer Science, Electrical, Civil, or Aerospace Engineering. For research-oriented or specialized roles, a Master's (M.Tech/M.Sc) or Ph.D. is often required. Successful candidates are placed in various ISRO centres across India, contributing to different stages of space missions—from design to launch.

Research and Development Careers

ISRO offers immense scope for those interested in research and innovation. The organization's premier research centres, such as the Vikram Sarabhai Space Centre (VSSC), U R Rao Satellite Centre (URSC), Space Applications Centre (SAC), and Liquid Propulsion Systems Centre (LPSC), provide opportunities for scientists and researchers to work on cutting-edge projects.

Positions like Junior Research Fellow (JRF), Research Associate (RA), and Project Scientist allow researchers to explore new technologies in propulsion, remote sensing, materials science, and satellite design. These posts generally require candidates to have completed an M.E./M.Tech or M.Sc. with NET or GATE qualification, while Ph.D. holders can apply for advanced research positions.

Administrative and Support Jobs

Apart from scientific work, ISRO also employs professionals in administrative, managerial, and technical support roles. These include positions in Accounts, Finance, Administration, Library Science, Purchase and Stores, as well as Medical Services. Each of these areas plays a crucial part in ensuring the smooth operation of ISRO's centres and missions.

Candidates with Diplomas, ITI certifications, or Graduate degrees in relevant disciplines are eligible for these positions. These roles may not directly involve rocket science, but they are vital to the organization's success and mission readiness.

Working with ISRO through Other Agencies

In addition to direct recruitment, there are opportunities to work with ISRO through its commercial and support agencies. Antrix Corporation, ISRO's commercial arm, handles satellite launches and marketing of space products. NewSpace India Limited (NSIL) focuses on the commercialization of space technologies. Moreover, ISRO collaborates with the Indian National Space Promotion and Authorization Centre (IN-SPACe), which supports and regulates private space startups.

Working with these organizations allows individuals to contribute to India's growing space ecosystem and gain valuable experience in space applications, commercialization, and innovation.

Career Growth and Promotions

ISRO offers excellent opportunities for career advancement and professional development. Scientists usually begin as Scientist/Engineer – SC and can move up to higher grades such as SD, SE, SF, SG, and beyond, depending on their experience and achievements.

The organization also encourages continuous learning through training programmes, research fellowships, and higher education sponsorships. Many ISRO scientists pursue doctoral studies or international collaborations, further strengthening their expertise and global exposure.

Salary and Benefits

ISRO provides competitive salaries along with a range of benefits. The starting pay for an entry-level Scientist/Engineer (SC) is approximately ₹56,100 per month, based on the 7th Pay Commission, with total monthly earnings ranging from ₹70,000 to ₹90,000 when allowances are added.

Employees also receive additional benefits such as housing facilities, medical care, pension schemes, travel concessions, and research grants. ISRO is known for maintaining a professional yet family-like work culture that values innovation and dedication.

Preparation and Eligibility

To succeed in ISRO recruitment exams, candidates must have strong fundamentals in Physics, Mathematics, and Engineering. Preparing with GATE-level concepts, practicing problem-solving, and studying previous ISRO exam papers are essential steps. Apart from technical knowledge, good analytical ability and creativity are key qualities that ISRO values in its scientists and engineers.

Alternative Space Career Paths

Even if one does not directly enter ISRO, there are several related career paths in India's space sector. Organizations like DRDO (Defence Research and Development Organisation), HAL (Hindustan Aeronautics Limited), BEL (Bharat Electronics Limited), and BHEL (Bharat Heavy Electricals Limited) frequently collaborate with ISRO and offer roles in aerospace technology and research.

In recent years, private space startups such as Skyroot Aerospace, Agnikul Cosmos, and Bellatrix Aerospace have also emerged, offering exciting new avenues for engineers and scientists. Additionally, pursuing research at institutions like IISc (Indian Institute of Science), PRL (Physical Research Laboratory), or IITs can also open doors to a career in space scien

Why ISRO?

Because a nation once looked up at the stars and refused to believe they were out of reach. ISRO was born from a dream — a dream to prove that even with limited means, unlimited vision could touch the heavens. It exists because India chose knowledge over doubt, courage over comfort, and unity over impossibility. Every mission, every rocket, every satellite is a reminder that science is not just about technology — it's about faith in human potential. ISRO stands as a symbol of India's relentless spirit: to explore, to innovate, and to inspire generations to look up and whisper, "We can."

ISRO @ 75: The Pride, Power, and Promise of India

Completing 75 glorious years, ISRO stands as a symbol of India's scientific spirit, determination, and innovation. From launching a humble sounding rocket to reaching the Moon and the Sun, ISRO has shown the world what vision, perseverance, and teamwork can achieve. Every mission tells a story of courage, dedication, and national pride. Congratulations to ISRO and all the brilliant minds who made India shine among the stars.

THE STORY OF ISRO

ISRO stands as a shining testament to India's unwavering spirit of exploration, innovation, and self-reliance. From its modest beginnings in a small church in Thumba to making global headlines with lunar and solar missions, ISRO's journey is nothing short of legendary.

Over the decades, ISRO has not only advanced space technology but also used it for the nation's development — connecting remote villages, improving weather forecasting, supporting agriculture, and enabling disaster management. Every satellite launch, every successful mission, and every scientific achievement represents the dedication, intellect, and teamwork of thousands of Indian scientists and engineers who dreamed big for their nation.

ISRO's success is not just about rockets or satellites — it's about the power of vision, unity, and national pride. It inspires every Indian to believe that no dream is too high, and no goal is beyond reach. As we celebrate 75 glorious years, we salute the heroes who turned imagination into reality and continue to guide India toward a brighter, space-powered future.

Congratulations, ISRO — 56 years of excellence, courage, and inspiration!

------ ★ -------